Climate Change Reconsidered

A New Look at What the Global Temperature Record Actually Shows

George Root

December, 2015

Climate Change Reconsidered

A New Look at What the Global Temperature Record Actually Shows

By George Root

December 2015: Second Edition

ISBN-13: 978-1519607386

ISBN-10: 1519607385

Dedication

This book is dedicated to Michael Crichton (1942-2008) whose book "State of Fear" introduced me to the Global Warming debate.

"Their beliefs will not be swayed by the facts for their beliefs do not depend upon facts, but rather on a deep seated need to believe."

– Carl Sagan

Table of Contents:

Preface: 6

1 - Introduction: 8
    1.1 - What is the Climate Debate Actually About?   8
    1.2 - A Word About the IPCC   9
    1.3 - A Word About the Word "Anomaly"   9

2 - Critique of the IPCC Climate Models: 10
    2.1 - There are several defects in the IPCC Climate Models: 10
    2.2 - But Surely a "Consensus of Scientists" Must Be Right?   12
    2.3 - But Surely Climate Scientists Wouldn't Lie?   13

3 - Analysis of the Global Temperature Data 14
    3.1 - Two Temperature Data Sets Are Used in this Study: 14
    3.2 - Fitting Analytic Functions to Data is a Backbone of Science   15
    3.3 - Fitting Analytic Functions is Part Art and Part Science   16
    3.4 - Cyclic Changes in the Earth's Climate   17
    3.5 - Cyclic Phenomena Tend to Follow Sinusoidal Oscillations   18
    3.6 – Outline of the Analysis to be Done   19

4 - Long Period Cyclic Changes in Climate: 20

5 - Cyclic Changes in Climate During the Modern Period:............. 22

6 - The Evidence for CO2 Warming 24

7 - The Final "Cyclic Climate Model" 28

8 – Analyzing "The Weather": 33

9 – Future Temperature Measurements:............................... 37

10 – Comparison with IPCC Predictions: 38

11 – Long Range Predictions: 42

12 – Final Thoughts: 43

References:... 44

Preface:

This is a small book – really just a research paper. But I believe that the contents may be some of the most important that you will read. That is because it provides answers to some really important questions that climate scientists have been unable to answer for decades. Questions upon whose answers the future of modern civilization may depend. OK, that's probably a little dramatic. Let's just say that the information contained in this little book is pretty important. It provides a completely different picture of Global Warming than the one you have been hearing from the IPCC (the "Intergovernmental Panel on Climate Change") and the "consensus" of climate scientists. Things may not be as dire as they would have us believe. There is a lot of evidence in the global temperature data that supports this conclusion and I will explain it in the following pages.

First, I'm not a climatologist. I'm an engineer. So I don't pretend to know what makes the climate tick, but I do know a lot about data analysis. And that's what I'm going to do here – analyze the global temperature data.

Data analysis uses the same techniques regardless of what the data represent. The data could be the daily closing price of a particular stock, or it could be the estimated annual temperature of the Earth. So, I'm going to use some standard data analysis tools to see if there are any patterns in the global temperature record that can shed light on some of the most difficult questions facing the climate community – and the rest of us:

1) How much of the recently observed temperature rise is due to natural causes?

2) Is there any hard evidence that CO_2 in the atmosphere is causing global warming?

3) How much of the observed warming since 1850 is due to CO_2?

4) Is global temperature going to increase without limit leading to a climate catastrophe by 2100?

5) What will the likely global temperature be in 2100?

So, as I said, those are some pretty important questions. For those of you who would prefer to just skip to the end of the book, here are the answers to those questions:

1) 87%

2) Yes

3) 13%

4) No

5) About 0.1 °C above the present

The fact that this analysis will supply these answers is, by itself, quite remarkable. The climate models used by the IPCC cannot do that. I will discuss the IPCC Climate Models a little later.

One of the important features of the analysis that will be presented here is that it makes relatively specific predictions about future global temperatures and these predictions differ greatly from those provided by the IPCC. So, it is going to be simple to compare what actually happens in the next few years with both sets of predictions and to decide, based on observed evidence, which is more correct.

1 - Introduction:

One of the great mysteries of the late 20th and early 21st centuries is the question of what is causing the observed warming in the Earth's climate. The figure below illustrates the estimated temperature data for the years from 1850 to 2014. The recent rise in temperatures certainly appears to be troubling if one assumes that that increase in temperatures is going to continue into the future. But is it?

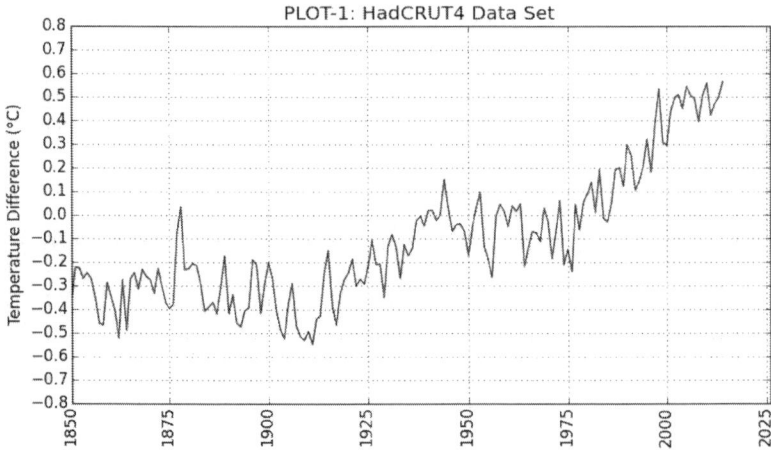

The purpose of this paper is to present a new analysis of the measured temperature data in order to resolve this question. Rather than trying to model the climate, an impossibly difficult task in light of the current state of knowledge, I will instead model the temperature data itself. If there are patterns in the temperature data, it may be possible to project those patterns into the future, even without knowing precisely what is causing those patterns to occur.

1.1 - What is the Climate Debate Actually About?

If you read or listen to the "mainstream media", you will probably believe that there is a "consensus of scientists" who believe that the Earth's climate is changing and a group of demented "deniers" who assert that the climate is not changing. Nothing could be further from the truth.

Actually, the "believers" and the "deniers" agree on a couple of points:

1) The Earth's climate is changing. It has always been changing. It will continue to change.

2) Climate changes before around 1950 were entirely due to natural causes. None of the changes before 1950 were due to CO_2.

The debate is about the period after 1950. The "believers" believe that essentially all climate change since then has been due to CO_2 in the atmosphere. The "deniers" believe that some or indeed most of the changes since 1950 have been a simple continuation of natural variations in the climate that have been going on for hundreds or thousands of years. In this paper, I will present an analysis that suggests that both of these views are at least partially correct.

Although this is a small book, it contains some "blockbuster" conclusions that refute the "climate catastrophe" predictions hypothesized by the IPCC. The most important of these conclusions are:

1) The climate is not a chaotic process as has been assumed, but is in fact composed of several entirely natural, cyclic, predictable components.

2) Almost all of the warming that has occurred since 1850 is due to a natural recovery from the "Little Ice Age" combined with other natural cyclic variations in climate.

3) There is evidence for an additional warming since 1950 that may be due to CO_2, but the maximum possible amplitude of this warming is +0.22 °C per century. Far below the dire predictions of the IPCC.

4) Global temperature in 2100 is very likely to be no more than 0.15 °C above the current level.

If you would like to know how I have arrived at these conclusions, read on.

1.2 - A Word About the IPCC

The IPCC is the "Intergovernmental Panel on Climate Change" formed under the auspices of the UN. Many climate scientists contribute to the reports published from time to time by the IPCC, but the IPCC itself is basically a political organization. I will use the term "IPCC" to refer to the organization itself as well as to all the climatologists who contribute to its publications.

1.3 - A Word About the Word "Anomaly"

"Anomaly" is a fine word in the English language. It has a definition which can be found in any dictionary: "an unexpected, unusual, or strange condition, situation, or quality". For some reason climatologists have chosen to use the term "Temperature Anomaly" when they mean to say "Temperature Difference". I will use the word "anomaly" to mean an unexpected data point or sequence of data points in the temperature data. It will turn out that are a few anomalies in the temperature record that provide useful information.

Following the convention used by the IPCC, the y-axis in all of the graphs in this book will show temperature difference from some unspecified baseline. This makes small temperature differences easier to visualize.

2 - Critique of the IPCC Climate Models:

So, why do we need another analysis of the global temperature record? Don't the IPCC climate models give us all the information we need to understand what's going on? Actually not.

The measured temperature data are the only actual evidence we have that the Earth's climate is warming. Computer models do not provide evidence. They simply produce whatever output they are programmed to produce. The IPCC started with the "CO2 Hypothesis" that atmospheric CO_2 was causing most or all of the observed temperature increase in the later half of the 20th and early 21st centuries. It is not surprising that the computer programs created by the IPCC support this hypothesis. Still, they do not provide any proof that this hypothesis is correct.

The analysis of the actual temperature data I will present in this book shows that the "CO2 Hypothesis" is, in fact, false. Not only is it false, it is impossible.

2.1 - There are several defects in the IPCC Climate Models:

1) They do not include any model of naturally occurring climate change. This is not surprising since there is currently very little knowledge about what has and is causing the Earth's climate to change naturally.

   To be honest, the IPCC computer models do include one possible natural cause of climate change. They allow that changes in the total heat output of the Sun will cause changes in the temperature of the Earth. However, it is well known that total solar output doesn't vary enough to account for the observed temperature change since 1950. So, the IPCC uses this weak natural effect to assert that "the observed temperature change cannot be explained by <u>any</u> natural causes".

   I will show that this statement is also false. Most of the observed temperature change since 1850 has been entirely due to natural causes.

2) They fail to predict what we know has already happened. Because the IPCC computer programs do not contain models for natural climate change, they cannot predict the natural variations in temperature that are known to have already occurred. The two biggest problems for the IPCC models are the cooling periods that have occurred from 1945 to 1975 and from 1998 until the present. Since atmospheric CO_2 was increasing during this entire period, it is difficult for the IPCC to explain why global temperatures did not also increase.

   I will show that both of these cooling periods are entirely natural and are not associated with CO_2 in any way.

3) They include a hypothetical "positive feedback" for which there is no physical justification. I have read that the IPCC climate models actually do

not predict a climate catastrophe due to CO2. They actually show a relatively small rise in temperatures due to CO2. In order to "pump up" the effect of CO2, the IPCC has reportedly invoked an unspecified "positive feedback" mechanism that, coincidently, multiplies the computed effect of CO2 by just enough to match the observed temperature increase. I have read that this positive feedback multiplier is on the order of 4 to 5. In other words, the IPCC has to multiply the computed effect of CO2 by a factor of 4 to 5 in order to predict the climate catastrophe it asserts is going to happen.

I will show that the effect of CO2 on global temperatures actually is quite small. Just as the un-fudged IPCC models reportedly predict it to be. In this way both the IPCC models and this analysis of the temperature data may actually agree.

4) They do not adequately model the effect of water vapor in the atmosphere. This is perhaps the biggest defect in the IPCC models.

The hypothetical "positive feedback" mechanism mentioned in the previous section is attributed to water vapor. The idea is that the small temperature increase due directly to CO2 will cause warming of the oceans and that will in turn result in more evaporation of water. It is known that water vapor is a powerful greenhouse gas and so, it does seem plausible that increased evaporation from the oceans might add additional warming to that caused directly by CO2.

Although this explanation might seem plausible, there are two problems with it.

a) The effect of water vapor in the atmosphere is very complex. If it remains as a gas, it does cause warming as the IPCC assumes. But that additional water vapor might also condense forming clouds. If low altitude clouds are formed, that leads to cooling by reflecting sunlight back to space. Or if the additional water vapor condenses as high altitude clouds, that would cause warming by trapping heat, preventing it from radiating to space. At present there is inadequate knowledge of the effect of water vapor to predict whether adding water vapor to the atmosphere would cause heating or cooling or both.

b) In order to "pump up" the small computed effect of CO2 on global temperatures, the IPCC has reportedly postulated a positive feedback. But systems with positive feedback tend to be unstable. It is one thing to assume that 1-unit of heating due to CO2 will be multiplied into 4-units of heating due to feedback. But those 4-units of additional heating would also be subject to positive feedback which would multiply their effect into 16-units of heating. And then 64-units. And so on until a "climate catastrophe" would result.

We are all familiar with this type of positive feedback. When a microphone that is attached to a sound system gets too close to the speakers, any small sound gets picked up by the microphone, amplified by the sound system, and broadcast from the speakers. The microphone picks up this amplified sound and sends it back through the amplifier where it gets even louder. And so on. The result is that the sound system puts out the loudest sound it is capable of producing.

Is it possible that the "tipping points" and "climate catastrophe" predicted by the IPCC are simply artifacts of the positive feedback assumption they have built into their models?

2.2 - But Surely a "Consensus of Scientists" Must Be Right?

Facts cannot be determined by voting. It doesn't make any difference how many scientists say something is true - scientific truth cannot be created by a consensus.

There have been at least three previous periods when the "consensus" of scientists believed an hypothesis that was later proven to be wrong:

1) **The Sun Revolves Around the Earth**: This was the "consensus" view of astronomers from the time of Ptolemy. It was not until Johannes Kepler successfully matched his elliptical orbit equations with the accurate observations of Tycho Brahe in 1609 that it was proven that the Ptolemaic model was wrong. We now know that this "consensus" of astronomers were wrong and that the Earth really does revolve around the Sun.

2) **Continental Drift Doesn't Exist:** I can remember reading in an early geography class text that a naive student might look at the left coast of Africa and the right coast of South America and think that they appear to fit together like the pieces of a puzzle. But a "consensus" of geologists know that this is simply a coincidence. There is no way those two coasts could ever have been joined. And yet, we now know that Continental Drift does exist.

3) **There Are No Genetic Differences Between Human Races:** This preposterous hypothesis was first proposed in the 1970s. I can remember reading about it in Scientific American. It was first proposed by a small group of "social scientists" in an attempt to stop any possible speculation that one race might be genetically superior to another. In other words, it was an attempt to be "politically correct". The politically correct pressure did succeed in suppressing any opposing papers until just recently when the ability to do rapid genetic sequencing became available. The first book proving the falsity of this hypothesis was published in 2014 ("A Troublesome Inheritance: Genes, Race and the Rise of the West" by Nicholas Wade). Genetic testing of tens of thousands of individuals around the globe proves without any remaining doubt that there are at least 5 and perhaps 7 genetically different "races" of humans: Africans, Caucasians, Asians, Australian Aborigines, and American Aborigines.

2.3 - But Surely Climate Scientists Wouldn't Lie?

Actually they would and have.

1) **The "Hockey Stick" Graph:** I'm sure that everyone has seen the famous "Hockey Stick" graph created by Michael Mann and published by the IPCC. This graph purports to show estimated global temperature over the past 1000 years based on studies of tree rings and such. It shows a basically stable temperature until the last half of the 20th century and then a rapid rise in temperature thereafter. It is quite convincing in demonstrating the global warming "catastrophe" that awaits us unless we do something quickly. The only problem is that it is pure fiction. At least two independent studies of the graph and the data leading to it have found that the "unusual" statistical analysis technique used by Mann was largely responsible for the shape of the curve. The IPCC has pulled this graph from its later publications.

2) **"An Inconvenient Truth"**: Al Gore's famous movie did not win him an Oscar, but it did win him a Nobel Peace Prize. But a problem arose in the UK. It turns out that it is not permitted in the UK to show propaganda films to school children. A group of skeptics took the Gore film to court claiming that there were at least 32 false or misleading statements made in the film and that it was therefore propaganda. The court agreed that large parts of the film are in fact, false or misleading. Now that really is inconvenient.

3) **"Climate Gate"**: In November of 2009 a series of email messages were leaked from the CRU (the Climatic Research Unit of the University of East Anglia in the UK), one of the four world data centers for climate research. These emails clearly showed that many of the most influential climate "scientists" have been manipulating data to make it appear that global warming is greater than it actually is, that they have been withholding data that tends to contradict the hypothesis that CO2 is a major cause of warming, and that they have been conspiring to discredit scientists who do not agree with them.

3 - Analysis of the Global Temperature Data

The climate is a very complex system. There are many inputs: solar radiation, cooling of the Earth's core, CO_2 and other greenhouse gases in the atmosphere, land and snow albedos, etc. The climate system mixes all these inputs together in ways mysterious and largely unknown and produces as a principle output, global temperature. Global temperature is the output of primary interest and has become synonymous with "climate". Although past temperatures are of some academic interest, it is prediction of future temperatures that is critical to the discussion of "climate change". Any approach that cannot predict future temperatures with some degree of accuracy is not of much value.

There are two ways to approach the problem of predicting future temperatures:

One way is to create a computer model of the climate making use of all available information about how the climate works to convert inputs into temperature. This is the approach taken by the IPCC and other climate scientists. As described in Chapter 2, this approach is hampered by incomplete or non-existent knowledge of how the climate actually works making it impossible to create a complete and accurate computer model. As mentioned in that chapter, the existing IPCC models do not predict, without post-facto fudging, temperature variations that are known to have already happened. This makes it very difficult to take their predictions of future temperatures seriously. This is particularly true since there are essentially no natural temperature changes included in their predictions.

An alternate approach is to start with the only hard data available, the measured or estimated past global temperature data and to determine whether there are any patterns in those data that might allow a prediction of future temperatures to be made. This is the approach taken in this book.

3.1 - Two Temperature Data Sets Are Used in this Study:

The Loehle Data Set: In order to find any long period oscillations in temperature, it is necessary to use a data set spanning several centuries. The "Loehle" data set includes estimated temperatures for the past 2000 years. These data were compiled by Craig Loehle from 18 different temperature reconstructions based on non-tree ring temperature proxies as created by various researchers from around the world. Loehle combined these temperature data into a single data set Ref (1). The portion of the "Loehle" data set spanning the years from 1500 up to 2000 was used in this study. Because I was unable to find a source for the original Loehle data, I digitized a published graph of the data on 25 year increments. These data are illustrated in PLOT-2 by the dotted curve.

The HadCRUT4 Data Set: In order to find short period oscillations in temperature, it is necessary to use a data set digitized on a short time increment. HadCRUT4 is a dataset of global historical land and sea surface temperatures spanning the years from 1850 up to the present (2014). This dataset is a collaborative product of the Met Office Hadley Centre and the Climatic Research Unit at the University of East Anglia. Ref (2). Although these data are available on monthly increments, the annual mean temperature data were used in this study.

The two temperature data sets used in this study are illustrated in PLOT-2. It is possible to draw one important conclusion from a simple examination of PLOT-2. The modern period warming illustrated by the HadCRUT4 temperature data appears to be a continuation of a warming trend that started around 1650 - around the time of the coldest period in the "Little Ice Age".

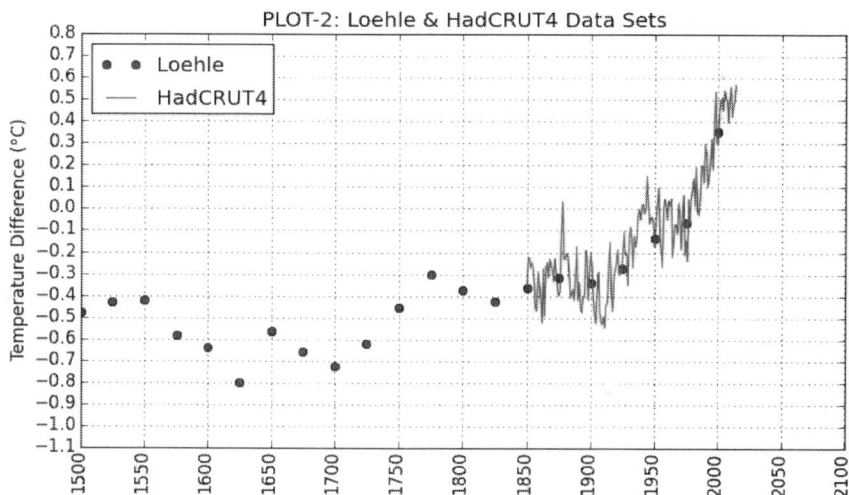

PLOT-2: Loehle & HadCRUT4 Data Sets

3.2 - Fitting Analytic Functions to Data is a Backbone of Science

The process of trying to identify analytic functions that correspond to measured experimental data has been a backbone of scientific progress for centuries. For millennia a "consensus of scientists" believed that the Earth was at the center of the Solar System. Only by comparing the observed motions of the planets (the data) with analytic functions predicted by Kepler's ellipses and Newton's Law of Gravity was it possible to prove otherwise.

In this paper, I will continue this tradition by fitting analytic functions to measured temperature data spanning the years from 1500 AD up to the present. I will use two types of analytic functions: sinusoidal oscillations representing cyclic changes in temperature and a residual linear trend that does not fit the

15

cyclic prediction after the year 1950. It is possible that this residual linear trend represents the contribution of CO_2 to the measured temperature variations.

3.3 - Fitting Analytic Functions is Part Art and Part Science

It is important to understand how curve fitting works. The "fitter" chooses the form of the analytic function s/he wants to try to fit to the data. That form is described by a mathematical equation. For example, a straight line function (linear regression) is defined by an equation of the form $Y = a*X + b$. In this example, "a" and "b" are unknown parameters whose values define a specific instance of the straight line and "X" and "Y" represent the data. Automated computer programs then sift through all possible versions of that specific shape and select the single set of parameter values, "a" and "b", that best fits the data. This computational portion of the process is exquisitely objective. The measure of how well a given form fits the data is quantified by the "mean square error" between the analytic function and the data. The chosen instance is that specific set of parameter values that results in the minimum possible error for that specific form.

It is important to note that the computational process does not change the data, nor can it change the form of the fit. The only variables are the values of the parameters. The form is chosen by the "fitter" based on experience and intuition. So, if the "fitter" has done a poor job of selecting the form of the function, no amount of computation will result in a "good" fit.

There are two tests to determine whether a given function is a "good" fit to the data. The first is objective. The mean square error between the data and the best-fit function must be small. The second test is more subjective. The best fit must be "reasonable". It must fit the data well and it must be physically plausible.

As an example, looking at the HadCRUT4 temperature data shown in PLOT-1, it occurs to me that these data have a shape similar to that of a 2nd degree polynomial: $Y = a*X^2 + b*X + c$ where a, b, and c are the unknown parameters. The HadCRUT data and the best possible 2nd degree polynomial fit are shown in PLOT-3.

Objectively this 2nd order polynomial is a pretty good fit. The error between the polynomial and the data is relatively small. So, is this really a good fit? No, it is a terrible fit. It has two fatal flaws. First, it predicts that global temperatures will rise upward toward infinity in the coming years. Although this might warm the hearts of the IPCC crowd, there is another fatal flaw. The 2nd degree polynomial also "predicts" that prior to 1850 global temperatures were dropping down from infinity. We know from the Loehle data that this is not what actually happened prior to 1850. So, the 2nd degree polynomial hypothesis is not physically reasonable even though it does fit the data pretty well.

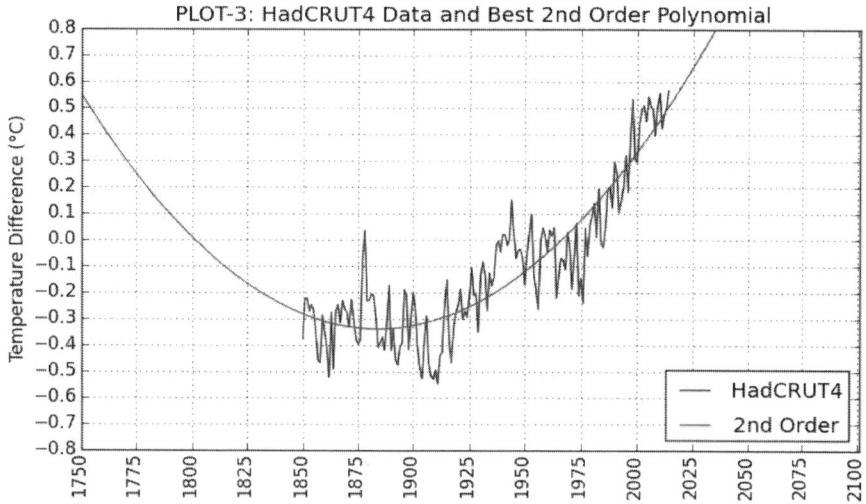

PLOT-3: HadCRUT4 Data and Best 2nd Order Polynomial

So, the first thing we need to do is to pick an analytic form that might fit the known temperature data and also that might be physically reasonable. It will turn out that a sinusoidal oscillation satisfies both of these criteria.

3.4 - Cyclic Changes in the Earth's Climate

It is known that there are cyclic changes in the Earth's climate that are entirely due to natural causes. In fact, it is apparent that there are cyclic changes in the temperature data shown in the graph of the HadCRUT4 data set in PLOT-1. Notice that the temperature increased from about 1910 to about 1945. Then it dropped until about 1975. Then it rose again until about 1998. Since then it has been fairly stable. This up-and-down pattern has a period of about 60 years.

There are several natural cyclic changes in the Earth's oceans and atmosphere that could cause cyclic changes in climate. Perhaps the best known of these are:

1) The "Pacific Decadal Oscillation" (PDO) which coincidently has a period of about 60 years,

2) The "El Nino Southern Oscillation" (ENSO), and

3) The "North Atlantic Oscillation" (NAO).

There is also an astronomical cycle which has a period of about 60 years. The major planets, Jupiter and Saturn, align on the same side of the sun about every 60 years. The gravitational pull of these planets causes tides on the sun, just as the gravitational pull of the Moon causes tides on Earth. How, or if, this astronomical cycle might be related to the roughly 60-year period of the PDO is unknown.

17

3.5 - Cyclic Phenomena Tend to Follow Sinusoidal Oscillations

The readily apparent cyclic variation in temperature illustrated in PLOT-1 combined with the known existence of cyclic changes in the Earth's oceans and atmosphere leads to the hypothesis that changes in the climate may also follow a cyclic pattern. The primary purpose of this study is to determine whether this hypothesis is correct, that is, is the measured temperature data consistent with a cyclic variation in climate?

Cyclic oscillations are characteristic of stable systems with negative feedback. The common household thermostat illustrates this. When the thermostat is set at a specific temperature, say 70° F, the room temperature does not go to 70° and stay there. It actually oscillates around 70°. As the room temperature drops, the furnace kicks in and adds heat causing the temperature to rise. As the temperature rises above the set point, the furnace stops and adds cooling by withholding heat. If the temperature is too low, the furnace adds heat. If the temperature is too high, the furnace adds cooling. The response is always opposite to the "forcing" – thus "negative" feedback.

This observation that stable systems often tend to oscillate around some "set point" is one reason for suspecting that the Earth's climate might also exhibit oscillatory behavior.

Naturally occurring cyclic changes such as the oscillations of a guitar string and ripples in a pond tend to follow a sinusoid pattern such as is illustrated in the figure on the next page. It therefore seems reasonable to look for cyclic changes in the temperature record which also follow a sinusoid pattern. The accuracy with which the sinusoidal hypothesis matches the temperature data will allow us to decide if the sinusoidal hypothesis is correct or not.

It is important to note that just because I will fit a sinusoidal pattern to the temperature data, that doesn't mean that the temperature data actually follows a sinusoidal pattern. I could try to fit any number of different patterns, such as the 2nd degree polynomial described above, to the data. Some would fit better than others. The conclusion that the temperature data is following a specific pattern depends entirely on how well the pattern fits the data. In this case, the sinusoidal pattern fits the data with remarkable accuracy. Sinusoids also have the necessary quality that they have bounded amplitude. They will never predict infinite temperature. They are physically plausible.

For our purposes, a specific sine wave is defined by the two parameters, Amplitude and Period as shown in the figure on the next page.

An Example of a Sinusoidal Variation in Temperature

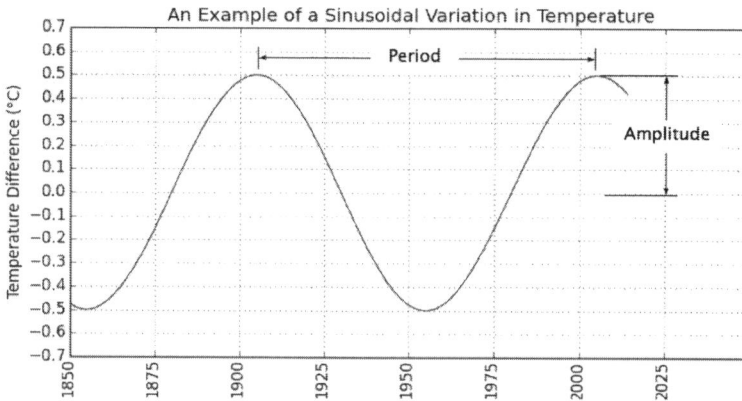

3.6 – Outline of the Analysis to be Done

Decomposing a complex data set into its analytic components is itself a cyclic process. We start with the raw data, the Loehle global temperature data set in this case. We compute the best sinusoidal fit to that data. The first fit will find the longest period sinusoid in the data (Sine.1). This fit will never be perfect. There will always be residual data that isn't well fit by the analytic function. We find this residual data by subtracting the best fit function from the input data. The difference is called "the residual data" or just "the residuals".

The "residual data" from the first fit becomes the input data for the second fit. Another best fit is found to the first residual data. This will be the next shorter period sinusoid (Sine.2).

The Loehle data I used are sampled on 25-year increments and this is too coarse to find any shorter period cycles, so at this point we have to switch to the HadCRUT4 data set. In this case, the input data are the residuals found by subtracting the first two sinusoids (Sine.12) from the HadCRUT4 raw data.

We go through the [fit -> subtract to find residuals] cycle twice to find two more sinusoidal oscillations in the HadCRUT4 data. At this point we will have found four sinusoids with periods of 1350-years and 247-years from the Loehle data and, 66-years and 9-years from the HadCRUT4 data.

The final residuals left after subtracting all four of the sinusoidal oscillations from the HadCRUT4 data have two components. Before 1950 the residuals are random year-to-year variations. There are no more analytic functions to be found in this portion of the data. However, since the final residuals are random, it is possible to apply statistical analysis tools to that data and set some limits on its expected behavior.

After 1950 there is a distinct pattern in the final residual data. This pattern is an upward linear trend that may be due to CO_2 warming.

4 - Long Period Cyclic Changes in Climate:

It is known from historical records that there is a climate cycle with a period of about 1000 years. It was warm during the "Medieval Warm Period" around 1000 AD, when the Vikings settled in Greenland, and again now, around 2000 AD in the "Modern Warm Period". Between these two warm periods, the climate cooled. The cool period between the "Medieval Warm Period" and the "Modern Warm Period" is known as the "Little Ice Age". In order to model this 1000-year cycle, we need to examine the Loehle temperature data extending back 500 years into the past.

PLOT-4 illustrates the Loehle data and the best long period sinusoidal fit to that data (Sine.1).

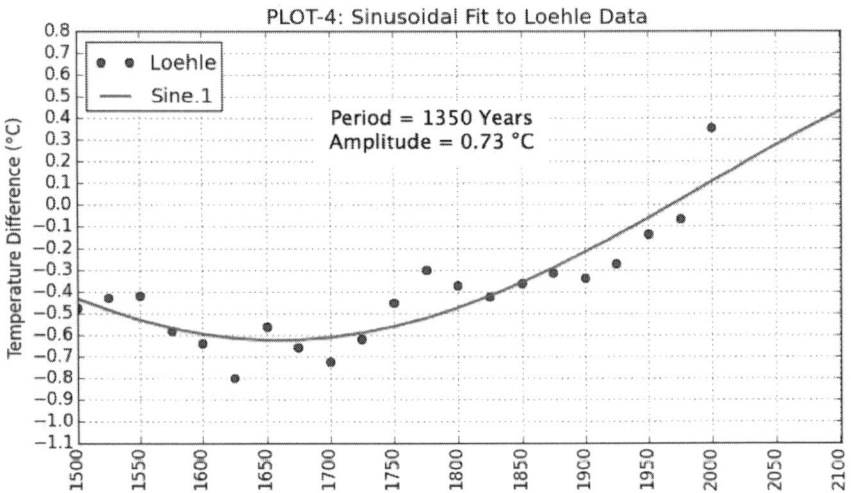

PLOT-4: Sinusoidal Fit to Loehle Data

Period = 1350 Years
Amplitude = 0.73 °C

The first conclusion is that this sinusoid, with a period of 1350 years, actually does fit the data rather well. No fit is ever perfect. A measure of how well the analytic function fits the data is the residual differences between the fit and the data. The residuals from the fit in PLOT-4 are shown in PLOT-5.

It is natural to ask, "Is there also a sinusoidal variation in the residuals shown in PLOT-5?" The way to answer that question is to try a sinusoidal fit and see how well it actually does fit the data. So, PLOT-5 shows not only the residual errors from the first fit, but also the best fit sinusoid (Sine.2) to these residuals. Once again, the analytic fit is remarkably good.

So, at this point we have decomposed the Loehle temperature data into two sinusoidal functions. One with a period of 1350 years and a smaller one with a period of 247 years. By adding these two sinusoids together, we arrive at a composite fit to the original Loehle data (Sine.12). This is shown in PLOT-6.

20

Sine.12 is a remarkably good fit to the Loehle temperature data. Particularly in the years following 1850 where the Loehle data overlap the HadCRUT4 data. We will examine this region in more detail in the next section. But even from this preliminary analysis, it is apparent that much of the warming since 1850 is a continuation of cyclic variations in climate that have been going on for at least 500 years.

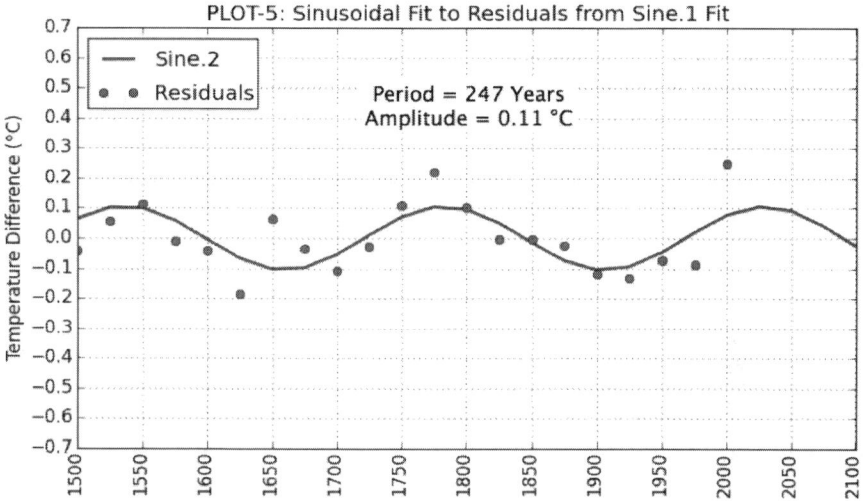

PLOT-5: Sinusoidal Fit to Residuals from Sine.1 Fit

Period = 247 Years
Amplitude = 0.11 °C

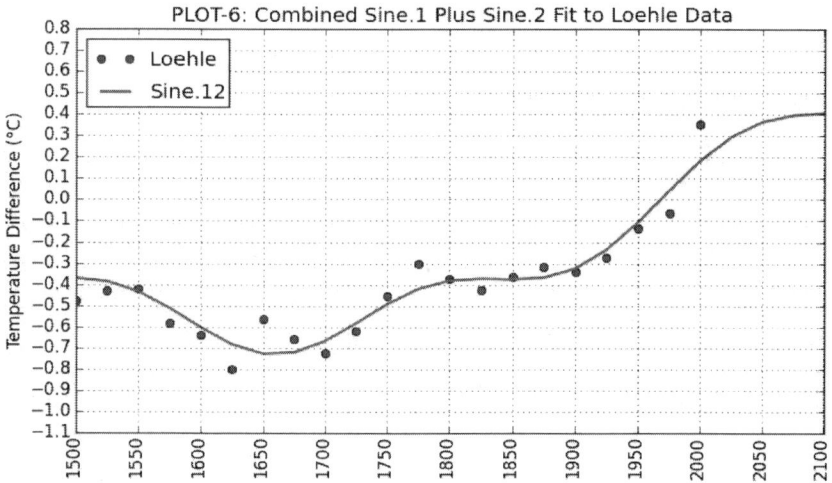

PLOT-6: Combined Sine.1 Plus Sine.2 Fit to Loehle Data

5 - Cyclic Changes in Climate During the Modern Period:

The long period cyclic variations in temperature derived using the Loehle data persist into the modern period. In order to determine whether there are additional cyclic variations during recent years, we must turn to the HadCRUT4 data set. PLOT-7 shows the HadCRUT4 temperature data along with the two sinusoid fit (Sine.12) from the Loehle data. Up to about 1990, the combined long period cyclic variation derived from the Loehle data also fit the HadCRUT4 data reasonably well. However, there are two significant differences between the HadCRUT4 data and the long term cyclic variation.

1) An additional cyclic variation in temperature with a period of about 60 years is readily apparent in PLOT-7 as it was in PLOT-1.

2) After about 1990 the HadCRUT4 data tend to rise above the long term cyclic variation. I will examine this in more detail later.

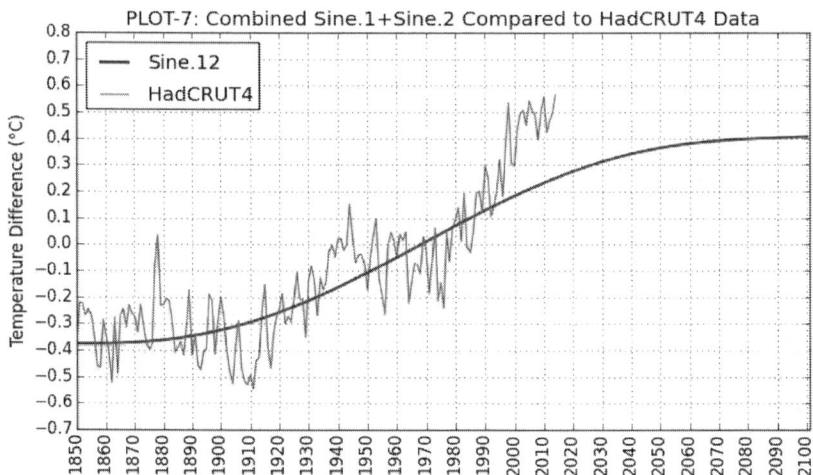

PLOT-7: Combined Sine.1+Sine.2 Compared to HadCRUT4 Data

In order to find and remove the 60-year cyclic variation in the HadCRUT4 data, we subtract the Loehle long period fits from the HadCRUT4 data resulting in the residuals not fitted by the long period sinusoids. These residuals are shown in PLOT-8 along with a new, short period, fit to those residuals. It is apparent that there is a relatively strong sinusoidal variation in temperatures during recent times. The period of this oscillation is 66 years.

Another interesting feature revealed in PLOT-8 is that there are sharp upward spikes in temperatures around 1880, 1945, and 1998 shown circled in PLOT-8. These spikes occur just before the climate starts to cool. It is beyond the scope of this study, and the author's knowledge, to try to find an explanation for this phenomenon but it would be a very interesting study. What causes these upward spikes? Is that cause also responsible for the periodic drops in temperature?

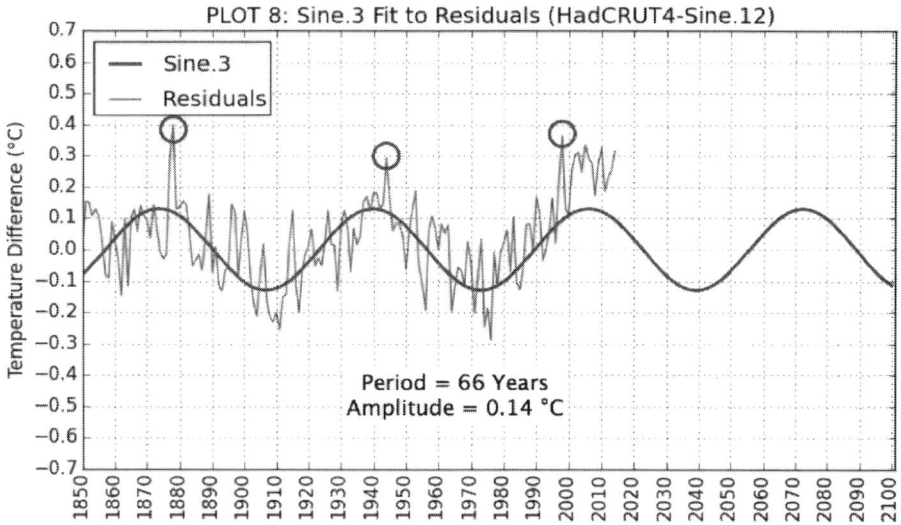

PLOT 8: Sine.3 Fit to Residuals (HadCRUT4-Sine.12)

Period = 66 Years
Amplitude = 0.14 °C

The combined three sinusoidal cyclic model (Sine.123) consisting of three sinusoidal oscillations in temperature with periods of 1350 years, 247 years, and 66 years is shown in PLOT-9 compared to the actual HadCRUT4 data.

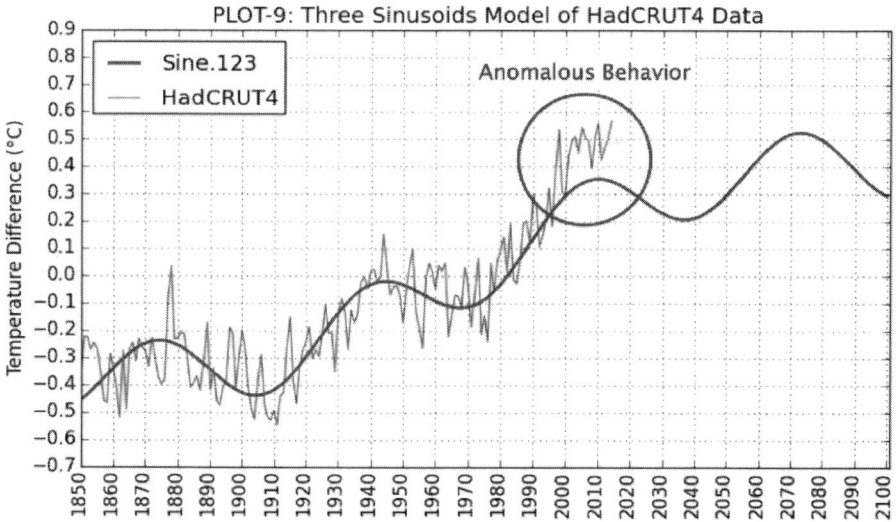

PLOT-9: Three Sinusoids Model of HadCRUT4 Data

Once again, the agreement between the data and the model is uncanny up to some time after 1950. After 1950, the data exhibit anomalous behavior in that they rise above the combined sinusoidal fit and remain there. This means that there is a component in the HadCRUT4 data after about 1950 that is not included in the three sinusoidal model.

23

6 - The Evidence for CO2 Warming

Subtracting the three sinusoids model (Sine.123) from the HadCRUT4 data results in the new residuals ("Residuals.4") illustrated in PLOT-10. These data represent the total error between the three sinusoids model and the HadCRUT4 data. Random errors would oscillate around "Temperature Difference" = 0.0. So, the behavior of the residual errors shown circled in PLOT-10 is anomalous because it rises above and stays above 0.0.

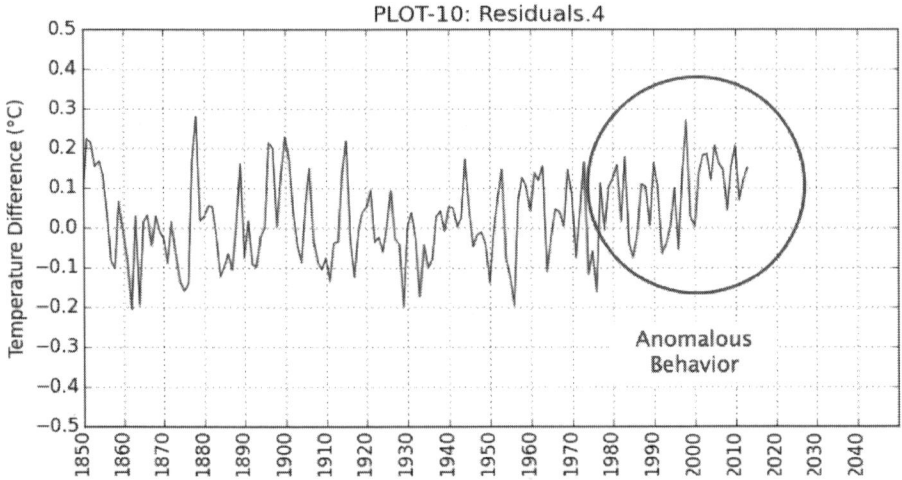

PLOT-10: Residuals.4

In order to proceed beyond this point, we must divide the Residuals.4 data into two portions, one containing the data prior to the anomalous data and the other containing the anomalous data itself. The problem is to determine exactly where the anomalous data begin. Choosing the correct starting point for the anomalous data is important because, ultimately, that will influence the amount of CO2 induced warming we will estimate. Choosing a starting year too early will tend to reduce the amount of CO2 warming and choosing a starting year that is too late will tend to inflate the amount of CO2 warming. So, how to determine where the anomalous data begin?

The problem is illustrated in PLOT-10a which simply repeats the plot of the Residual.4 data shown in PLOT-10 but without the circle around the anomalous data. Looking at the years 2000 to 2010, the anomalous behavior is obvious. But exactly where it begins is not at all obvious.

One possible way to make the transition from "normal" to anomalous data is to compute the "running sum" of the Residuals.4 data. To compute the running sum, we start at 1850 and simply add each succeeding year's temperature to the running sum. For random data, the running sum will stay near 0.0 since the positive excursions will tend to be cancelled out by the negative excursions. But, when we reach the anomalous data, the running sum will begin to become more and more positive.

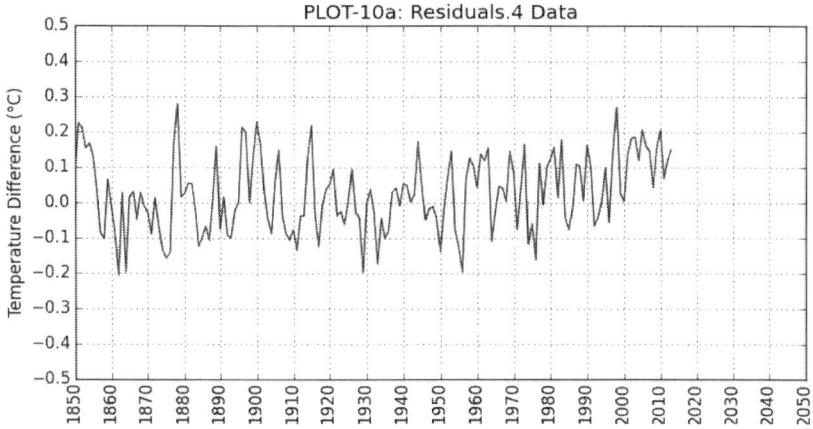

PLOT-10a: Residuals.4 Data

The running sum of the Residuals.4 data is shown in PLOT-10b. If it is assumed that this upward departure from the normal data is due to a linear trend in temperature, then the running sum should have the shape of a 2nd degree polynomial.

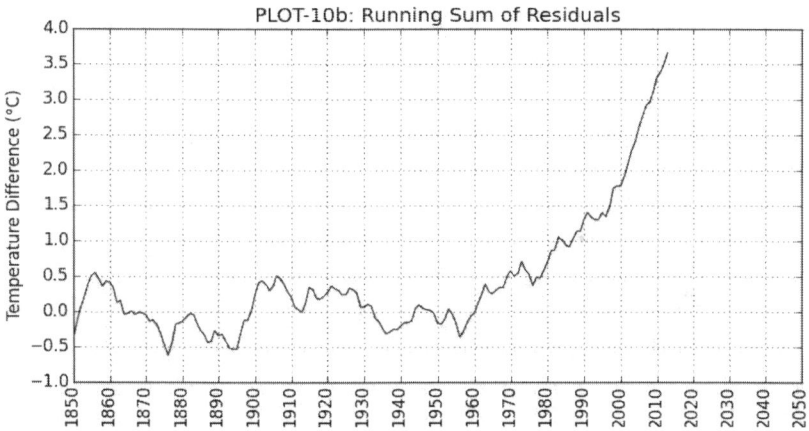

PLOT-10b: Running Sum of Residuals

PLOT-10c shows the results of fitting a 2nd degree polynomial to the running sum data starting at the year 1950. As can be seen, the 2nd degree polynomial does fit the anomalous data very well. This successful fit provides two pieces of important information about the anomalous data:

1) The anomalous data do in fact represent an upward linear trend in the Residuals.4 data. Else a 2nd degree polynomial would not have fit the running sum data.

2) The parameter values of the 2nd degree fit also supply the parameters for the linear trend itself. Basically, the running sum is the discrete integral of the Residual.4 data. Thus the linear trend that caused the anomalous

data is the derivative of the 2nd order polynomial. I know that this is all a lot of gibberish to many readers, but stick with me.

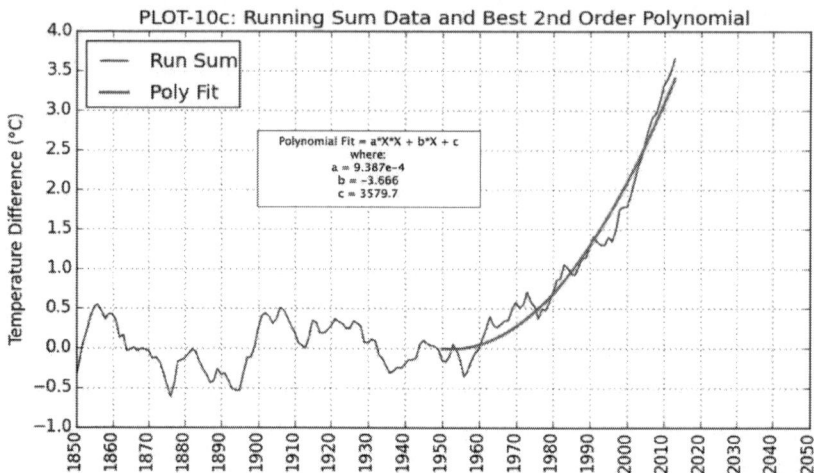

The 2nd degree polynomial shown in PLOT-10c is described by an equation of the form: Temperature = a*Year^2 + b*Year + c where the parameters "a", "b", and "c" have the values shown in PLOT-10c. The linear trend causing this running sum is then given by the equation:

Temperature = 2*a*Year + b.

Using the values for "a" and "b" from PLOT-10c we find:

Linear Trend in Anomalous Data = 0.188 °C per Century
Linear Trend Starts at Year 1950

This confirms the assertion that, prior to 1950, all variations in global temperatures were due entirely to natural (cyclic) causes. After 1950 there is an upward linear trend in the data that may be due to CO_2.

Now that we have confirmed that 1950 is indeed the starting year for the upward linear trend, we can use a second technique to find the parameters for this trend. We can simply fit a linear trend to the Residuals.4 data after 1950. The result of this exercise is shown in PLOT-11 on the next page.

Note that the estimated trend derived from a direct linear fit to the Residuals.4 data (0.23 °C/century) is slightly different from that derived by differentiating the 2nd degree polynomial fit to the running sum (0.19°C/century). I will use the 0.23 °C/century value since it allows a slightly larger warming component. This might very well be the "signature" of CO_2 induced warming that has been so illusive to the IPCC.

26

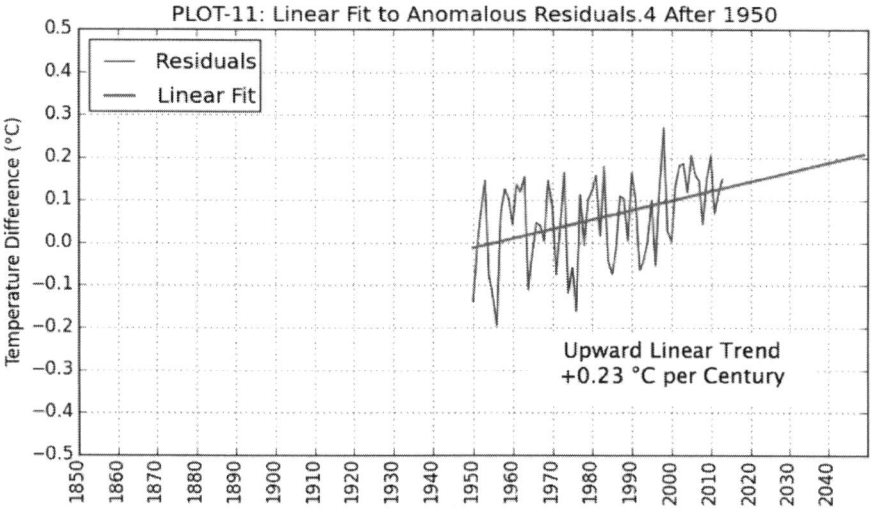

PLOT-11: Linear Fit to Anomalous Residuals.4 After 1950

But, what about prior to 1950? PLOT-12 shows the results of fitting a sinusoidal variation to the Residuals.4 data prior to 1950. The best fit sinusoid to these data has a period of 9 years. But the amplitude is so low, 0.042 °C, that it is questionable whether this is a real cyclic variation in temperature or just a best fit to random noise. I will include this 4th sinusoidal variation in the final "Cyclic Climate Model" to be compiled later simply to avoid leaving out something that might later prove to be important.

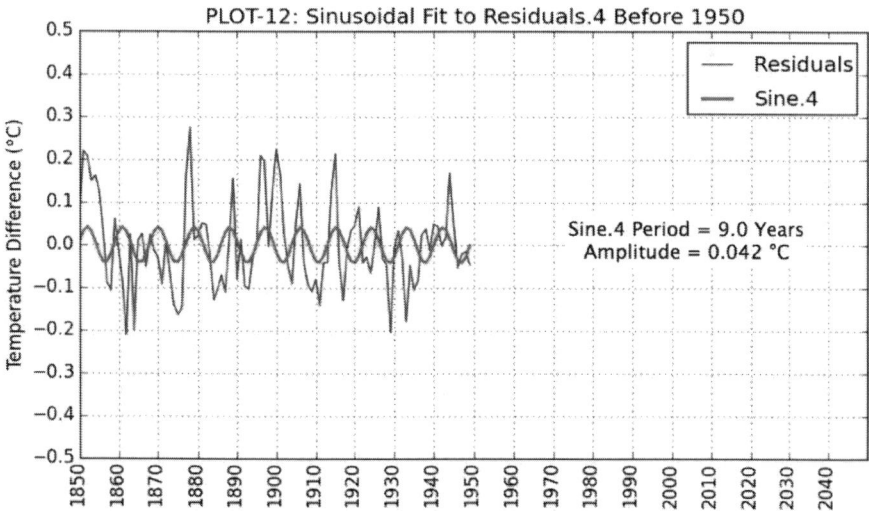

PLOT-12: Sinusoidal Fit to Residuals.4 Before 1950

7 - The Final "Cyclic Climate Model"

At this point I have decomposed the HadCRUT4 temperature data into a "Cyclic Climate Model" with five components:

> **Sine.1**: Period = 1350 years, Amplitude = 0.725 °C
>
> **Sine.2**: Period =   247 years, Amplitude = 0.106 °C
>
> **Sine.3**: Period =   66 years, Amplitude = 0.130 °C
>
> **Sine.4**: Period =   9 years, Amplitude = 0.042 °C
>
> **Linear Trend** after 1950 = 0.23 °C/century

PLOT-13 shows each of these components separately. Notice that Sine.1 which is the 1350-year period sinusoid is, in this range of years, essentially a linear trend. But a linear trend would not be a plausible fit to this temperature data for the reasons stated previously: a linear trend implies infinite temperatures in the future. The Sine.1 component is amplitude limited to ±0.725°C.

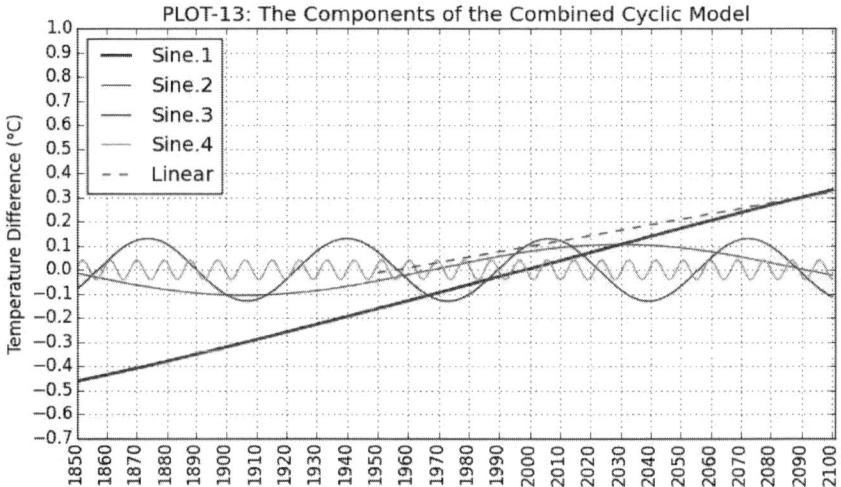

PLOT-13: The Components of the Combined Cyclic Model

Combining these five components results in the final "Cyclic Climate Model" illustrated in PLOT-14. I have included the +0.23 °C per century upward trend after 1950 in PLOT-14 to account for possible CO2 warming. PLOT-14 also shows the HadCRUT4 temperature data. Personally, I am dumbstruck at how closely the "Cyclic Climate Model" fits the data. It shouldn't even be possible to fit climate data at all if climate really was chaotic and to find such a simple model that matches the general trends in climate over the past 164 years so accurately is remarkable.

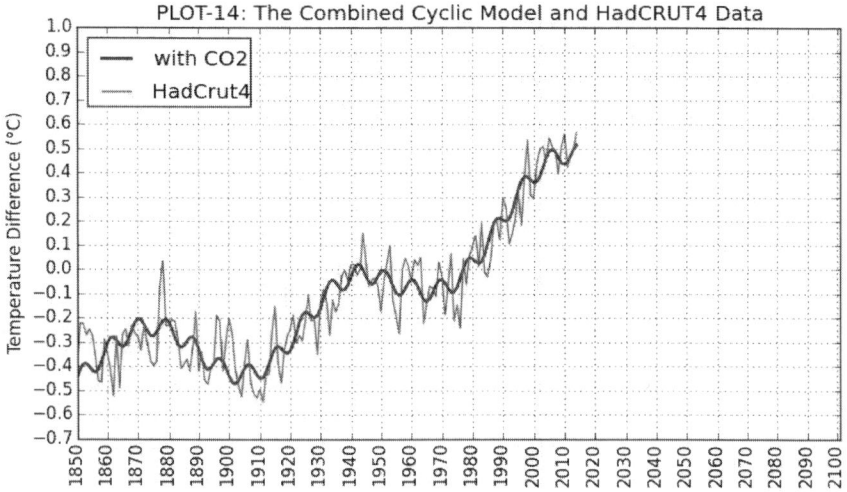

PLOT-14: The Combined Cyclic Model and HadCRUT4 Data

PLOT-14 shows the close match between the "Cyclic Climate Model" and the HadCRUT4 temperature data. Up until this page in this book, it has been assumed that climate is a "chaotic" process. A chaotic process is one that has three characteristics:

1) It is driven by numerous uncorrelated random processes with numerous feedback mechanisms.

2) The final outcome, global temperature in this case, is extremely sensitive to the state of the climate at any instant. Even a slight change in that state can cause large changes in climate in coming years.

3) And most important, chaotic processes cannot be predicted. Weather is recognized as a chaotic process. It is impossible to predict the weather more than 5-10 days into the future. It is not "difficult", it is "impossible". So, it has been natural to assume that climate is also a chaotic process and therefore unpredictable. This analysis proves otherwise.

If global temperatures were actually the product of a chaotic process, any attempt to fit the temperature data with any analytic function would have failed. The success of fitting the temperature data with such a simple model proves that climate, at least as represented by global temperatures, is not a chaotic process.

The "Cyclic Climate Model" permits the expected global temperature to be computed for any year in the range 1850 to 2014. The climate changes described by the "Cyclic Climate Model" have been driven by powerful, natural, cyclic forces for the past 164 years and there is no plausible explanation for why these powerful forces should suddenly stop in 2015. Although, this analysis doesn't prove that it will be possible to predict global temperatures into the future, it is very likely that that is possible. Only time will tell.

I know that you are probably tired of hearing about residuals by this time, but I do have to show you one more plot.

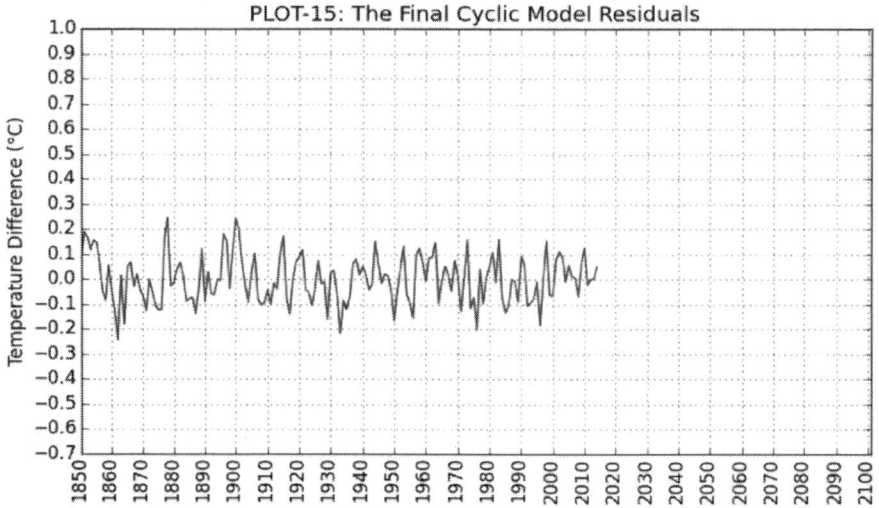

PLOT-15: The Final Cyclic Model Residuals

PLOT-15 shows the residual errors between the final "Cyclic Climate Model" and the HadCRUT data that are shown in PLOT-14. These residual errors appear to be random. I will prove that they are random a little later. Random year-to-year temperature variations are known as "weather".

At this point in the book I have demonstrated that the measured global temperature data (HadCRUT4) can be decomposed into three components:

1) **"Natural Climate"** - Consisting of the long term (decadal or longer) cyclic oscillations in global temperature that have been going on during the entire span of the HadCRUT4 temperature record (1850 up to the present (2014)) and probably far longer than that. These oscillations are entirely natural; they have nothing to do with CO_2 since they started long before 1950. There is no plausible reason that these cyclic oscillations should suddenly stop in 2015. These cyclic oscillations can be described by a fairly simple equation meaning that "Natural Climate" is predictable.

2) **"CO_2"** – There is a systematic deviation from "Natural Climate" in the temperature record for the years since 1950. This component can be modelled as an upward trend with a slope of about +0.23°C per century. This component is also completely predictable. It should be noted that there is no evidence in the temperature record that proves this component is due to CO_2, but that is a plausible hypothesis.

3) **"Weather"** – There are month-to-month and year-to-year random variations in the measured temperature record known as "Weather". This "Weather" component is random and is therefore not predictable. Although not predictable, the Weather component has a limited amplitude which can be characterized by a simple statistical analysis that I will present a little later.

The "Cyclic Climate Model" consists of the sum of the two analytic components of the temperature record: "Natural Climate" + "CO2". This "Cyclic Climate Model" can be expressed by a fairly simple equation consisting of four sinusoidal oscillations and a linear trend. Given any year, the equation can be solved for the "expected" global temperature for that year. The actual temperature for that year will be the "expected" temperature ± the "Weather".

It is apparent, and very fortuitous, that the random variability of global temperatures is small compared to the non-random, predictable, component. We can predict the large changes in temperature, but not the smaller, random changes.

And, finally, PLOT-16 shows the final composite "Cyclic Climate Model", both with the linear trend, possibly due to CO2, as well as the entirely natural variation in climate without the CO2 component. This is an extremely powerful figure from which it is possible to derive answers to several questions that have been frustrating the IPCC modelers for many years.

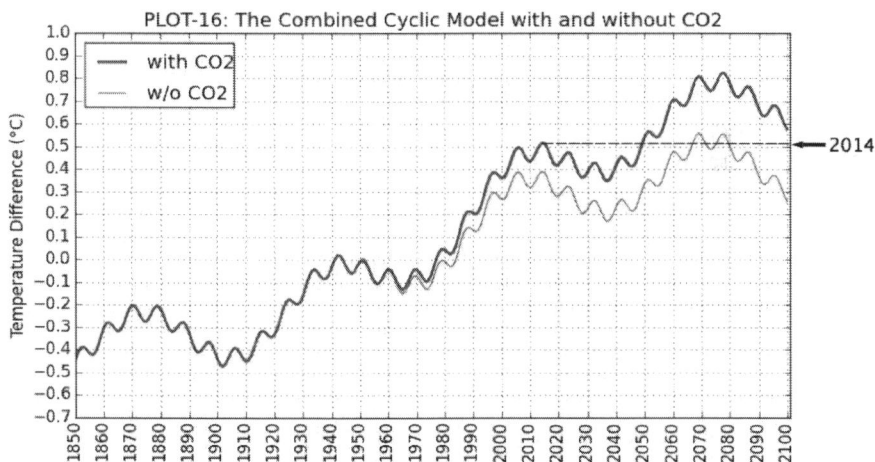

PLOT-16: The Combined Cyclic Model with and without CO2

For convenience, I have indicated the temperature at the end of 2014, which was slightly above 0.5 °C. Remember, everything to the left of 2014 is fact. The HadCRUT4 temperatures <u>do</u> follow regular cyclic behavior. There can be no dispute that this is true. PLOT-14 proves that it is true. Everything to the right of 2014 is prediction. But it is prediction based on solid evidence that temperatures

31

were predictable prior to 2014 and there is no plausible reason to suspect that the cyclic model will suddenly stop operating in 2015.

So, what can we learn from PLOT-16? Wonderful things:

1) The year 2015 will likely be "the hottest year on record" due to natural oscillations in the "Cyclic Climate Model"

2) The recent warming documented by the HadCRUT4 temperature record is almost entirely due to a natural recovery from the "Little Ice Age" which reached its minimum temperature around 1650. The Earth has been warming ever since then due to natural, but unknown, causes

3) The "pause" in global warming since 1998 is entirely due to natural cyclic variations in climate. This "pause" will turn into a general decline in global temperatures until about 2035. By 2035 the average temperature will be roughly 0.15°C below current temperatures.

4) After 2035, temperatures will again rise to the next "hottest year on record" in 2080. This will be about 0.35 °C above current temperatures.

5) Global temperatures have risen about 0.96°C since 1850. Of this rise, about 87% (0.83°C) has been due entirely to natural cyclic variations in climate.

6) In 2100, global temperatures will likely be about 0.1°C above current levels assuming that the CO_2 warming trend remains at 0.23°C/century.

7) In 2100, global temperatures will likely be about 0.3°C above the natural level due to CO_2 warming assuming that the CO_2 warming trend remains at 0.23°C/century

8) All of these predictions of future warming are far below those postulated by the IPCC. I will discuss this fully in a later chapter.

There is, to my knowledge, no explanation as to what physical processes may be causing the observed cyclic variations in temperature. However, the fact that we don't know what is causing the observed oscillations in temperature in no way invalidates the existence of such oscillations. The fact that these oscillations exist is clear in the temperature record itself.

8 – Analyzing "The Weather":

It is important to understand that we cannot measure the components of global temperature separately. All we can measure is the sum of all three components: "Natural Climate" + "CO2" + "Weather". PLOT-17 shows the measured temperature (HadCRUT4) which is the sum of all three components and the "Cyclic Climate Model" consisting of the sum of the two predictable components, "Natural Climate" + "CO2".

It is apparent in PLOT-17 that the "Cyclic Climate Model" fits the measured temperature data very well indeed. It is also apparent that the measured temperature hardly ever falls exactly on the predicted "Model". The measured temperatures generally fall above or below the expected "Model" temperature in a random fashion.

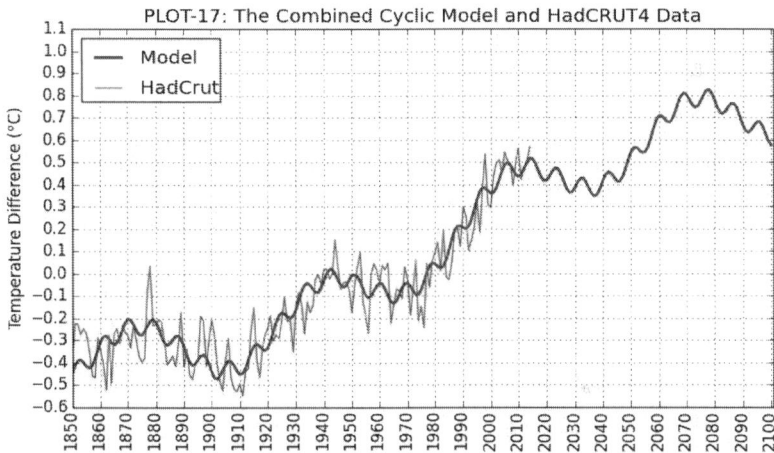

PLOT-17: The Combined Cyclic Model and HadCRUT4 Data

We can separate out the random "Weather" component by subtracting the predictable "Cyclic Climate Model" from the measured temperatures. The resulting "Weather" component is shown in PLOT-18. Although the "Weather" component is not predictable, it is possible to set some limits on its behavior.

One way to characterize random data is to compute a histogram of that data. Basically a histogram counts the number of times the value of the random data falls within a small range of amplitudes. In order to create a histogram of the "Weather" data, I divided the range of amplitudes from -0.3°C to +0.3°C into 11 bins. Each bin holds the count of the number of times the data falls within the range of that bin. The resulting histogram is shown in PLOT-19.

Also shown in PLOT-19 is a "Gaussian" fit to the histogram data. The Gaussian curve in PLOT-19 is an example of the familiar "bell shaped curve" which is common in statistical analysis of random data. In fact, the Gaussian distribution is so common, that it is also called the "Normal Distribution" and random data

33

are said to be "normally distributed" if they have a Gaussian distribution. It is apparent from PLOT-19 that the "Weather" data are "normally distributed". We can infer from this fact alone that "Weather" is likely due to a number of uncorrelated random processes.

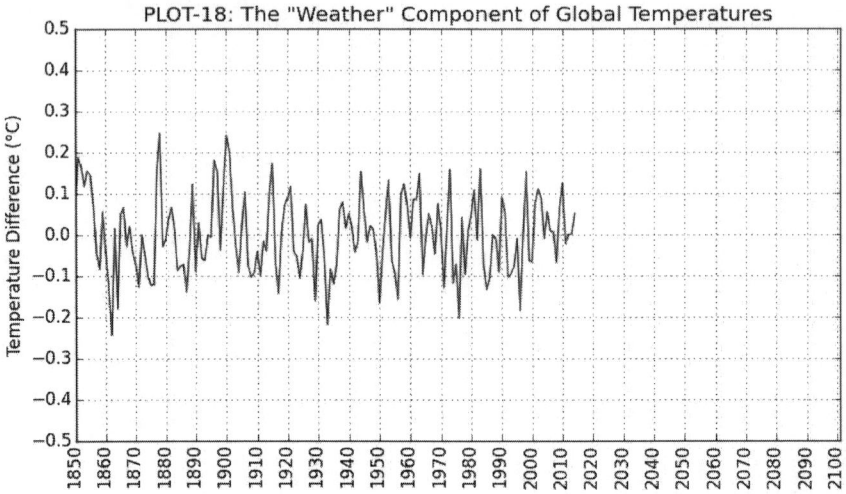

PLOT-18: The "Weather" Component of Global Temperatures

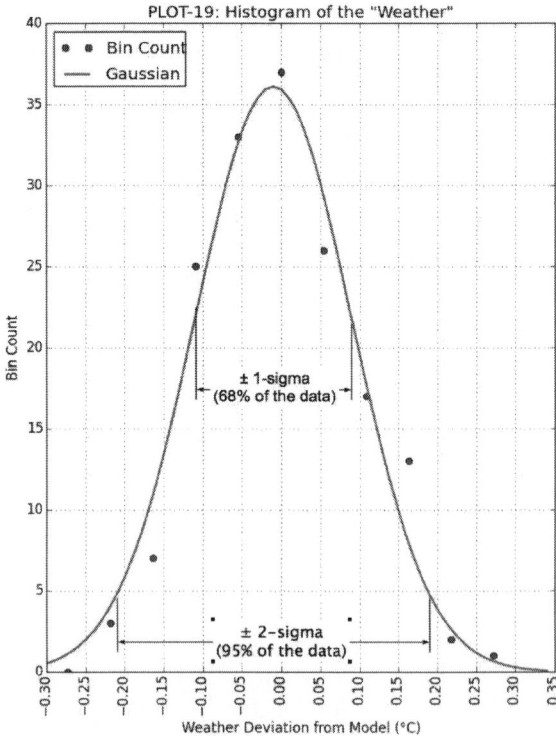

PLOT-19: Histogram of the "Weather"

The width of a normal distribution is characterized by a parameter called "sigma". This parameter is a measure of the "width" of the Gaussian distribution. So a large sigma means that the random "Weather" data vary widely around the expected "Cyclic Climate Model" predictions. A small value of sigma means that the measured temperature data tend to fall very close to the "Model" predictions. The 1-sigma value of the Gaussian distribution shown in PLOT-19 is 0.1°C.

OK, this can be confusing. The parameter "sigma" measures ½ the full width of the distribution. So it is customary to state that the "Model" predicts the actual measured temperature within ±0.1°C. The "±" doubles the sigma value and is a measure of the full width of the normal distribution. This is shown in PLOT-19.

Because the "Weather" is normally distributed, we can make some statements about the likelihood of future temperature measurements. For normally distributed random variables, roughly 68% of all measurements will fall within ±1-sigma of the "Cyclic Climate Model" temperature predictions. And, roughly 95% of all measurements will fall within ±2-sigma of the "Model" temperature predictions. Roughly 2.5% of all measurements will be more than 2-sigma (+0.2°C) above the "Model" predictions. These observations are very important for interpretation of future temperature measurements.

Before talking about future temperature measurements, let's check on those already made. For the HadCRUT4 data, there are 164 annual temperature measurements. Based on the normal distribution, we would expect that roughly 2.5% = 4 of those measurements would be higher than +0.2°C (2-sigma) above the "Model" predictions. There are actually only 3 years when the measured temperatures were +0.2°C or more above the "Model" predictions. So, we might expect to find more "hot" years in the future.

Another interesting characteristic of random phenomena is the "auto-correlation function". This is a measure of how much the weather in one year influences the weather in succeeding years. PLOT-20 shows the auto-correlation of the "Weather" component of global temperature shown in PLOT-18.

In PLOT-20 a Y-value of 1.0 means perfect correlation and a Y-value of 0.0 means completely uncorrelated. So, PLOT-20 shows that the "Weather" in one year is 30% correlated with the weather in the previous year. A hot year is more likely to be followed by another hot year than simple random chance. After 2 years, the correlation drops to zero meaning that the weather in a given year is completely uncorrelated with the weather two or more years previous.

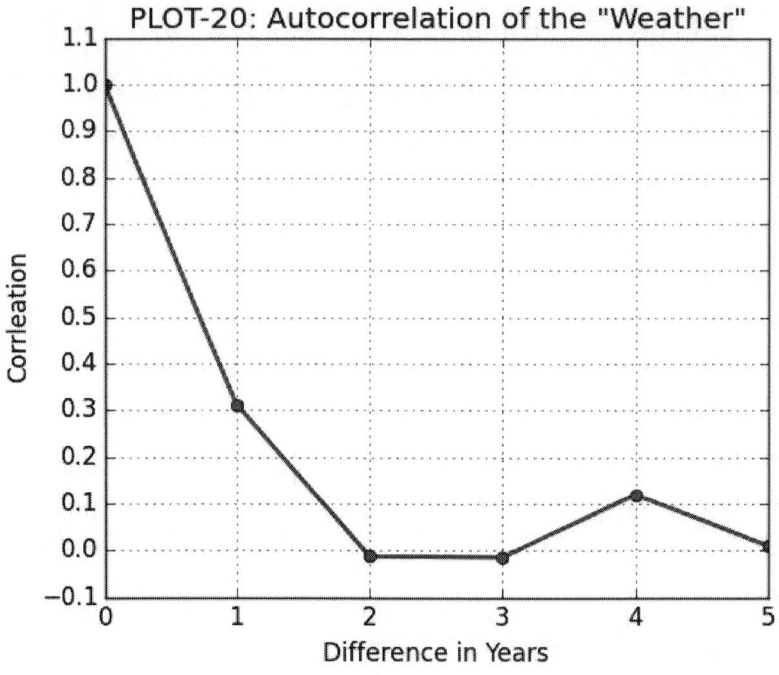

PLOT-20: Autocorrelation of the "Weather"

9 – Future Temperature Measurements:

The HadCRUT data for 2015 will be available in a month to two. Based on temperature data already available, 2015 is likely to be the "hottest year on record". This is not surprising for two reasons:

1) The "Cyclic Climate Model" predicts that 2015 will be the "hottest year on record" based solely on the natural cyclic oscillations in temperature combined with the 0.13°C of heating attributable to CO_2 in 2015. This is shown by the arrow in PLOT-21.

2) There is a very strong El-Nino weather phenomenon building in 2015 that is expected to persist into 2016. El-Nino events are associated with higher than normal global temperatures. It is important to note that El-Nino events are "Weather" events and have nothing to do with CO_2 in the atmosphere.

So, it is likely that 2015 and perhaps 2016 will have temperatures higher than the "Model" predictions. PLOT-21 shows the "Model" predictions of global temperature and ±2-sigma lines located ±0.2°C above and below the "Model". Once again, it is expected that 95% of future temperatures will lie between these ±2-sigma lines and roughly 2.5% of future temperatures will likely lie above the +2-sigma line. It will be interesting to see what 2015 and 2016 temperatures turn out to be.

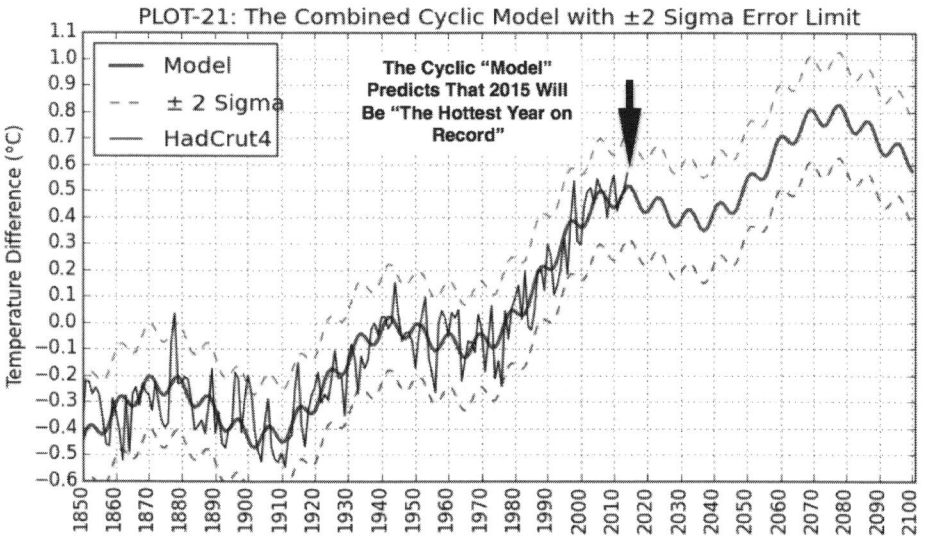

PLOT-21: The Combined Cyclic Model with ±2 Sigma Error Limit

10 – Comparison with IPCC Predictions:

It is natural to ask how the temperature predictions of the "Cyclic Climate Model" compare with the predictions of the IPCC computer models. One problem is that there is not a single "IPCC Prediction". There are myriad different computer models of the climate and each of these produces different forecasts. In addition, the IPCC has defined four "Representative Concentration Pathways" abbreviated to "RCPs" (Ref 3). The RCPs define standard sets of scenario assumptions about how CO_2 concentration, and other important inputs to the climate programs will change in future years. There are four standard RCPs:

1) RCP2.6 = CO_2 peaks in 2020
2) RCP4.5 = CO_2 peaks in 2045
3) RCP6.0 = CO_2 peaks in 2080
4) RCP8.5 = "Business as Usual" scenario

Projected temperatures for RCPs 4.5 through 8.5 are shown in PLOT-22 (Ref 4). Presumably RCP2.6 was not scary enough to be included. Actually RCP2.6 is, according to the IPCC, the only scenario that meets the goal of keeping temperatures below 2°C but it's assumption that CO_2 production will peak in 2020 is quite unrealistic.

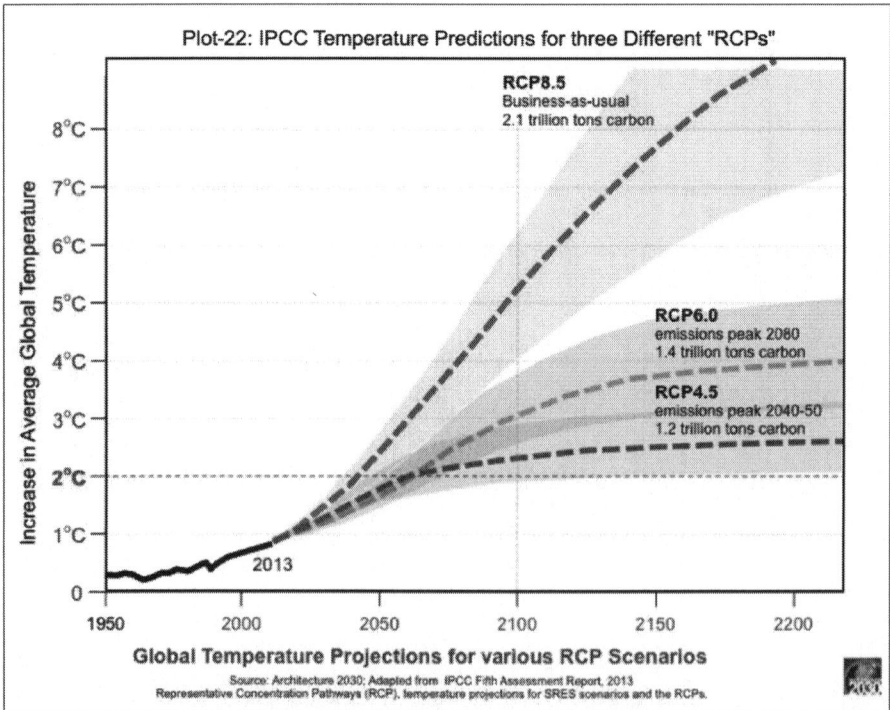

Plot-22: IPCC Temperature Predictions for three Different "RCPs"

RCP8.5
Business-as-usual
2.1 trillion tons carbon

RCP6.0
emissions peak 2080
1.4 trillion tons carbon

RCP4.5
emissions peak 2040-50
1.2 trillion tons carbon

Global Temperature Projections for various RCP Scenarios

Source: Architecture 2030; Adapted from IPCC Fifth Assessment Report, 2013
Representative Concentration Pathways (RCP), temperature projections for SRES scenarios and the RCPs.

There is wide variability between the four RCPs as would be expected from different assumptions about how much CO_2 will be in the atmosphere in coming years. Within each RCP there is also wide variability between the various climate model computer programs. This variability is indicated in PLOT-22 by the light gray shading around each projected curve. The uncertainty in the projected temperatures gets quite large after about 2050.

The horizontal dashed line at 2°C in PLOT-22 is significant for two reasons:

1) This is the temperature above which, according to the IPCC, "climate catastrophe" will occur. And second,

2) The 2015 United Nations Climate Change Conference in Paris (COP21) has agreed to set a goal of limiting global warming to less than 2°C. The conference attendees also agreed to "pursue efforts" to limit the temperature increase to 1.5 °C.

Please note that the "zero" baseline in PLOT-22 is not the same as that used for the HadCRUT4 temperature measurement data and the "Cyclic Climate Model" so, in order to make a direct comparison of the two temperature models, I had to adjust the RCP baseline to agree with the HadCRUT4 baseline. The result is shown in PLOT-23 for RCP4.5. RCP4.5 has the lowest projected temperatures of all the realistic RCPs so basically the IPCC is predicting a temperature at or above the RCP4.5 line in PLOT-23 (± the variability in the computed results).

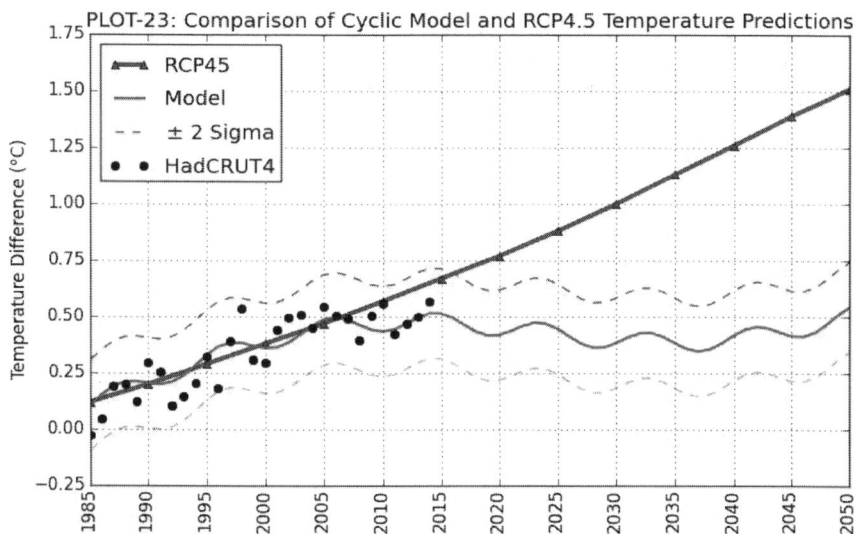

PLOT-23: Comparison of Cyclic Model and RCP4.5 Temperature Predictions

The first thing to notice in PLOT-23 is that the "Cyclic Climate Model" predicts that the IPCC goal of keeping global temperature rise below 2°C should be relatively easy to achieve. In fact, the global temperature rise will be below 1°C

39

at least until 2050 assuming that warming due to CO2 continues to increase at its current rate.

It is apparent in PLOT-23 that both the IPCC RCP4.5 model and the "Cyclic Climate Model" accurately "predict" measured temperatures prior to about 2005. This is not surprising since the data from before 2005 were available when both models were created and both models were created by fitting curves to the measured data.

The two models of future temperatures do begin to diverge after 2005. However, frustratingly, the rate of divergence is small enough to make a clear distinction between them difficult. Both lie within the margin of error of the "true temperatures" (HadCRUT4). Because it is expected that 2015 is going to be "the hottest year on record", for reasons previously discussed, it is unlikely that the data for 2015 will resolve the issue. Depending upon how much of the El-Nino warming carries over into 2016, that year may also not provide a clear indication of which model is being followed by the true temperatures. This is a classic example of fitting experimental data to analytic models in order to determine whether or not those models correctly represent the data.

So, the global temperatures predicted by the "Cyclic Climate Model" and the IPCC models are pretty similar right up to 2015. However, the contribution attributed to CO2 is remarkably different between those two models. The CO2 contributions to "Global Warming" for the two models are shown in PLOT-24.

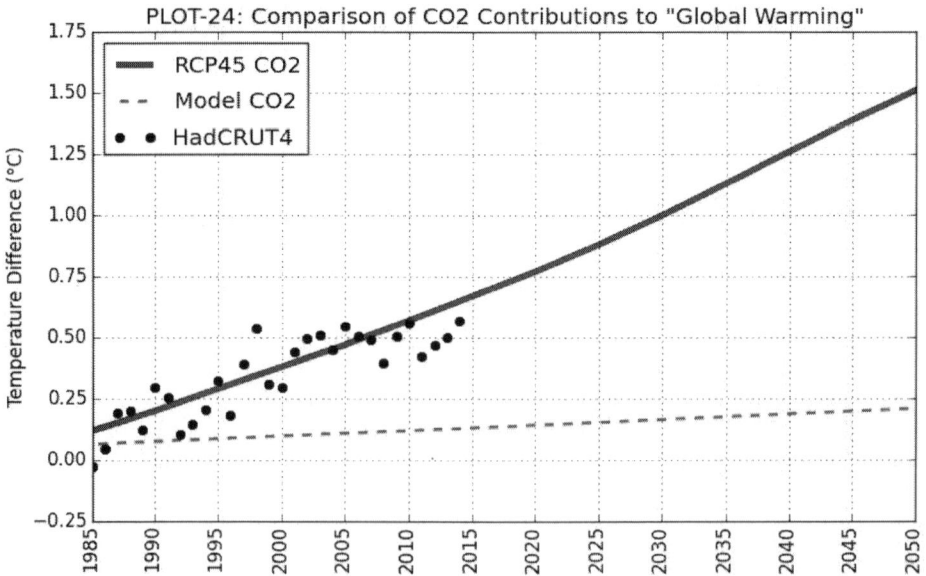

PLOT-24: Comparison of CO2 Contributions to "Global Warming"

The "Cyclic Climate Model", which has been shown to accurately predict global temperatures within ±0.1°C assigns the temperature increase from 1850 to 2014 as follows:

- Warming Due to Natural Cyclic Oscillations = 0.832 °C (87%)
- Warming Due to CO_2 since 1950 = 0.129 °C (13%)
- Total Warming 1850-2014 = 0.961 °C
- HadCRUT4 Temperature Increase = 1.013 °C

The HadCRUT4 temperature increase for 1850 to 2014 is 1.013 °C. The warming due entirely to natural causes is 0.832 °C. Thus, it is impossible for the contribution due to CO_2 to be more than 0.18 °C. The "Cyclic Climate Model" warming attributed to CO_2 is 0.13 °C leaving the remaining 0.05 °C in the HasCRUT4 data to "Weather".

The IPCC models are based on the hypothesis that all of the observed warming since 1950 has been due entirely to CO_2. Thus the total temperature and the temperature contribution due to CO_2 are identical for the IPCC models as indicted in PLOT-24. This result is impossible. This impossible hypothesis is caused by two defects in the IPCC models:

First, those models contain no natural climate drivers. Thus any increase in global temperatures must be due to CO_2.

Second, it has been reported that the IPCC models contain a positive feedback factor that multiplies the effect of CO_2 so that the computed temperatures match the observed temperatures. The IPCC predicted CO_2 contribution to temperature for 2014 is 0.67°C. The predicted 2014 temperature contribution due to CO_2 for the "Cyclic Climate Model" is 0.13°C. Taking the ratio of these two CO_2 contributions yields 4.8. Thus, the "Cyclic Climate Model" tends to confirm the IPCC's use of a positive feedback factor in the range 4X to 5X to "inflate" the contribution of CO_2 to global warming.

11 – Long Range Predictions:

The "Cyclic Climate Model" contains a sinusoidal component with a period of 1350-years. This makes it possible to "predict" temperatures well into the future. PLOT-25 shows three of the sinusoidal components of the "Cyclic Climate Model". The CO_2 component, also shown in PLOT-25, will be a significant contributor to "global warming" in coming centuries. In roughly 500 years it might contribute as much as $1.2°C$ of additional warming.

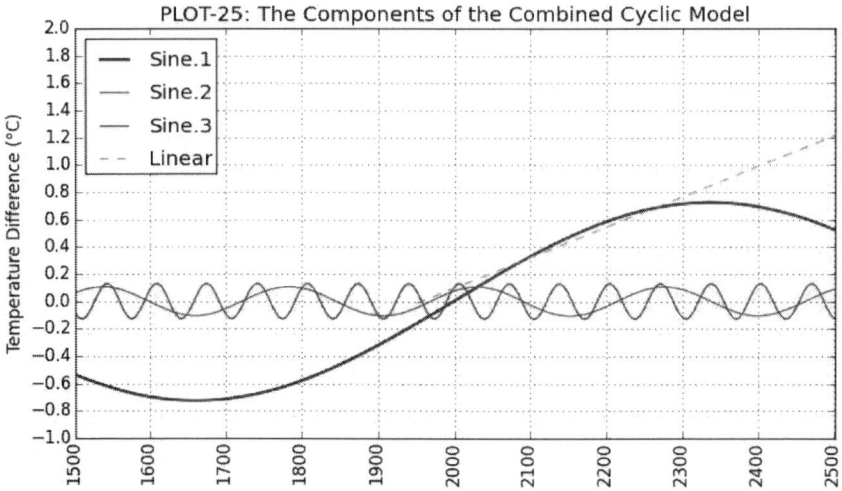

PLOT-25: The Components of the Combined Cyclic Model

PLOT-26 shows the combined "Cyclic Climate Model" projected out 500 years. Curves with and without the CO_2 contribution are shown. Even with the CO_2 warming trend at its current level, global temperatures in 2500 are still likely to be below the $2°C$ " climate catastrophe" proposed by the IPCC.

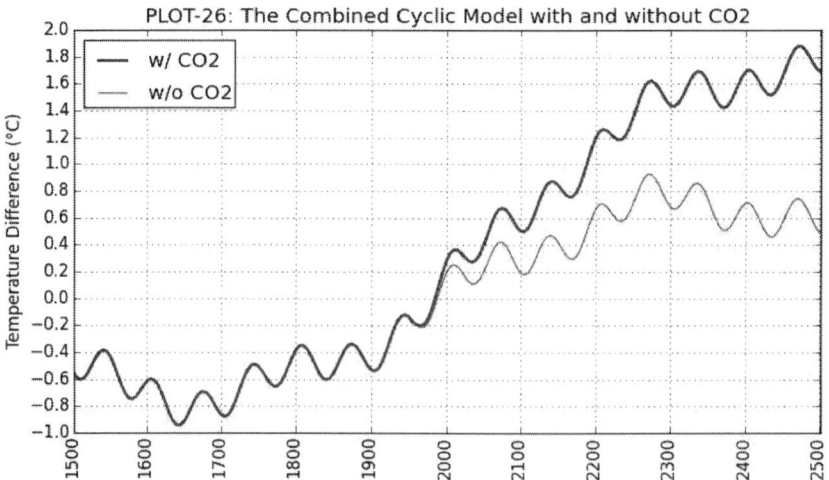

PLOT-26: The Combined Cyclic Model with and without CO2

12 – Final Thoughts:

This project started out as an exercise in learning to use the Python programming language. Python is easy to learn and a lot of fun to use. And it's free! All of the calculations in this book were accomplished using Python3 and its many numerical (NumPy), scientific (SciPy), and plotting (MatPlotLib) algorithms. I highly recommend it to anyone who wants to write computer programs for engineering and scientific studies.

When I started this study, I thought, as do many people, that global temperatures were generated by chaotic climate processes and were, therefore, unpredictable. But, as the study documented in this small book demonstrates, global temperatures are not random. The major changes in global temperature, as described by the "Cyclic Climate Model", follow a well defined analytic function.

The ramifications of this discovery are enormous both for the Earth's climate and for the IPCC climatologists and policy makers who believe that they can change the climate by reducing CO_2. They cannot. Global Warming is being driven almost entirely by powerful natural forces having nothing to do with CO_2.

Some may scoff and say that this study is nothing but "curve fitting". That is certainly true. It is also irrelevant. It doesn't matter how the "Cyclic Climate Model" was derived. It only matters that it does in fact accurately predict global temperatures.

It will probably be a disappointment to the climatologists who use the world/s most powerful computers to estimate future global temperatures to find that it is possible to predict expected global temperature with an accuracy of ± 0.1°C using the scientific calculator app on any cell phone.

OK – enough bravado. Here is the culmination of this study:

The "Cyclic Climate Model" Hypothesis:

There is no credible reason that the "Cyclic Climate Model" behavior that has been going on since 1850 will stop in 2015. It is therefore hypothesized that the analytic "Cyclic Climate Model" that accurately predicts global temperatures before 2015 will continue to do so after 2015. Fitting future measured temperatures to the "Cyclic Climate Model" predictions will determine whether this hypothesis is true or not.

References:

1) Craig Loehle, Energy & Environment - vol 18, No. 7+8, 2007, page 1052 ff. The series used were: GRIP borehole 18O temperature (Dahl-Jensen et al., 1998); Conroy Lake pollen (Gajewski, 1988); Chesapeake Bay Mg/Ca (Cronin et al., 2003); Sargasso Sea 18O (Keigwin, 1996); Caribbean Sea 18O (Nyberg et al., 2002); Lake Tsuolbmajavri diatoms (Korhola et al., 2000); Shihua Cave layer thickness (Tan et al., 2003); China composite (Yang et al., 2002) which does use tree ring width for two out of the eight series that are averaged to get the composite, or 1.4% of the total data input to the mean computed below; speleothem data from a South African cave (Holmgren et al., 1999); SST variations (warm season) off West Africa (deMenocal et al., 2000); SST from the southeast Atlantic (Farmer et al., 2005); SST reconstruction in the Norwegian Sea (Calvo et al., 2002); SST from two cores in the western tropical Pacific (Stott et al., 2004); mean temperature for North America based on pollen profiles (Viau et al., 2006); a phenology-based reconstruction from China (Ge et al., 2003); annual mean SST for northern Pacific site SSDP-102 (Latitude 34.9530, Longitude 128.8810) from Kim et al. (2004); and Spannagel Cave (Central Alps) stalagmite oxygen isotope data (Mangini et al., 2005). This gave a total of eighteen series with quite wide geographic coverage (including tropical) and based on multiple proxies.

2) P. Brohan, J.J. Kennedy, I. Harris, S.F.B. Tett and P.D. Jones, "Uncertainty estimates in regional and global observed temperature changes: a new dataset from 1850". J. Geophys. Res, 111, D12106, doi:10.1029/2005JD006548.

3) "The Beginner's Guide to Representative Concentration Pathways",

   http://www.skepticalscience.com/rcp.php?t=1

4) Architecture 2030 – Adapted from the IPCC 5th Assessment Report 2013.

   http://architecture2030.org/enews/news_111213.html

Printed in Great Britain
by Amazon